I0797330

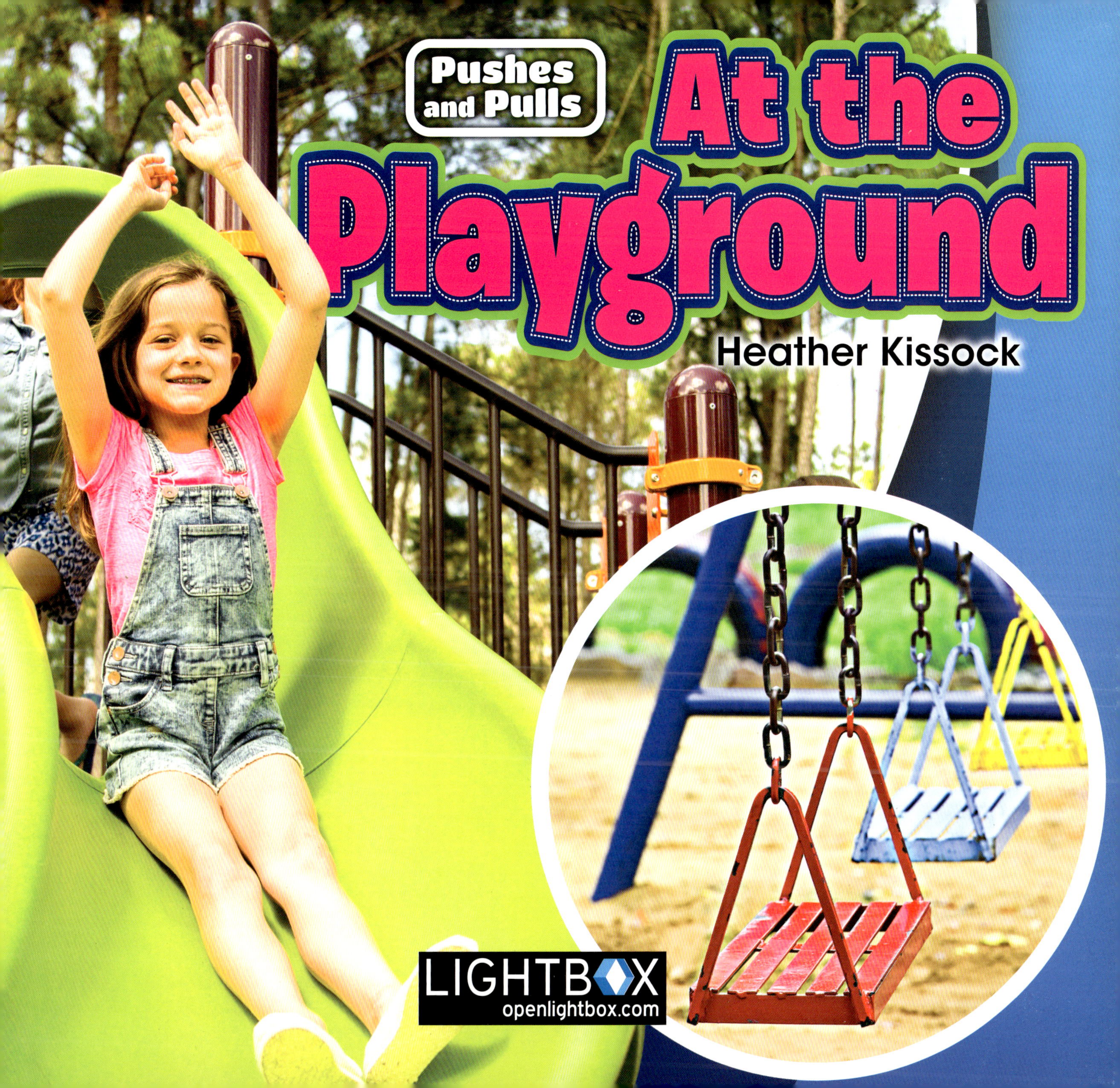
Pushes and Pulls
At the Playground
Heather Kissock
LIGHTBOX
openlightbox.com

Go to
www.openlightbox.com
and enter this book's
unique code.

ACCESS CODE

LBXU5299

Lightbox is an all-inclusive digital solution for the teaching and learning of curriculum topics in an original, groundbreaking way. Lightbox is based on National Curriculum Standards.

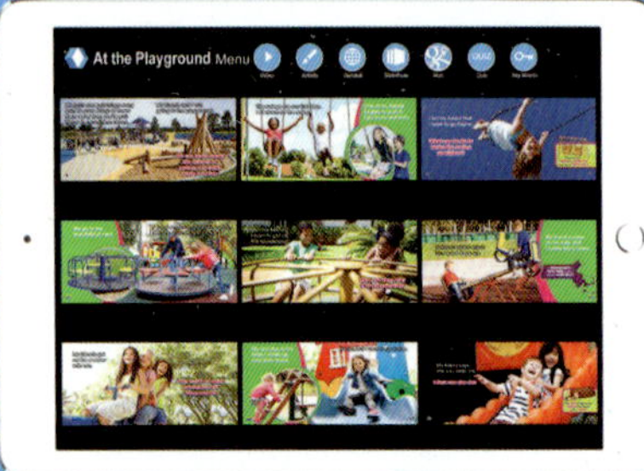

OPTIMIZED FOR

- ✓ **TABLETS**
- ✓ **WHITEBOARDS**
- ✓ **COMPUTERS**
- ✓ **AND MUCH MORE!**

STANDARD FEATURES OF LIGHTBOX

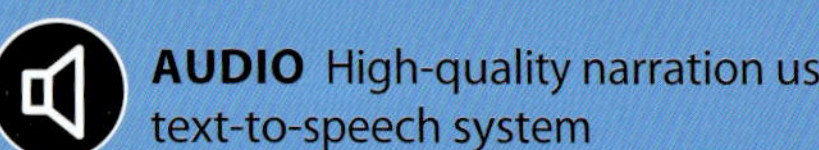
AUDIO High-quality narration using text-to-speech system

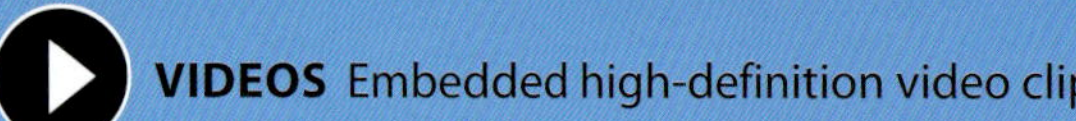
VIDEOS Embedded high-definition video clips

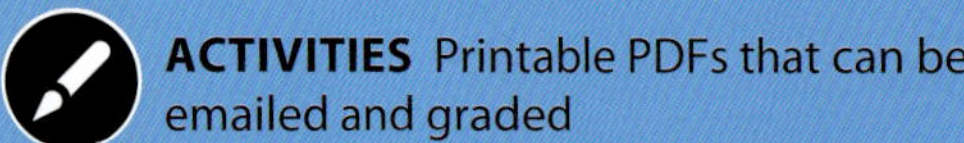
ACTIVITIES Printable PDFs that can be emailed and graded

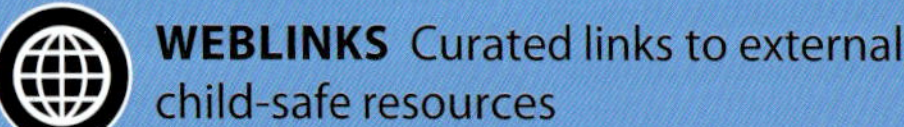
WEBLINKS Curated links to external, child-safe resources

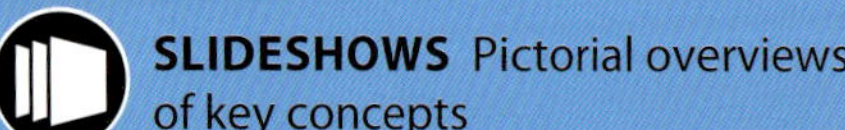
SLIDESHOWS Pictorial overviews of key concepts

INTERACTIVE MAPS Interactive maps and aerial satellite imagery

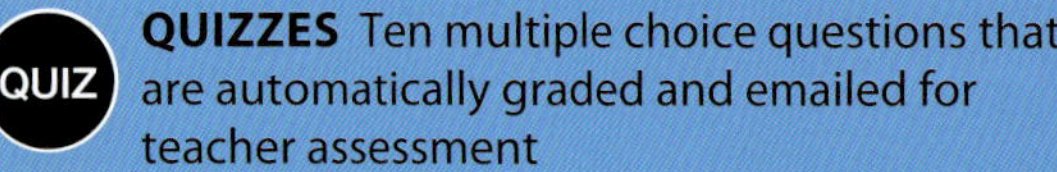
QUIZZES Ten multiple choice questions that are automatically graded and emailed for teacher assessment

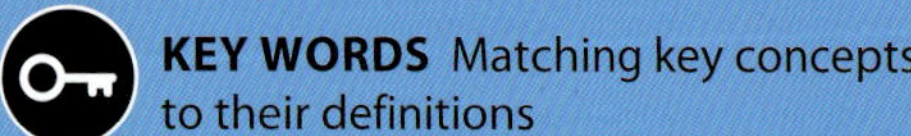
KEY WORDS Matching key concepts to their definitions

VIDEOS

WEBLINKS

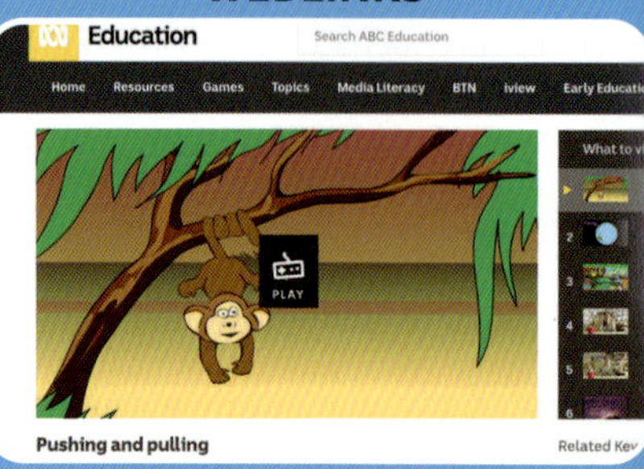

SLIDESHOWS

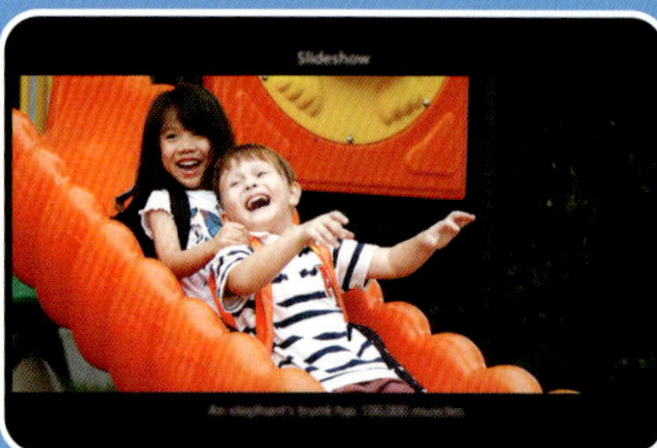

QUIZZES

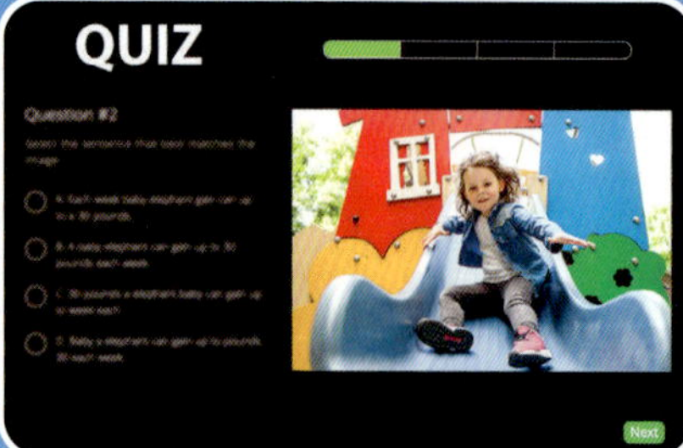

Contents

We push and pull things every day. We push things to move them away from us. We pull things to move them closer.

My friends and I are going to the playground.

We can learn about the science of pushes and pulls while we play.

The swings are our first stop.
I sit down on the swing.

One of my friends begins to push it. I go back and forth.

I tell my friend that I want to go higher.

What can he do to make the swing go higher?

People have **played** on **swing sets** for at least **3,500 years**.

We go to the roundabout next.

I jump on, and my friends push it.
It spins around, and so do I.

I tell my friends
I want to get off
the roundabout.

How can they stop it from spinning?

Seesaws are so much fun. I push to go up.

My friend pushes on his side, and I come back down.

In **Massachusetts**, a **seesaw** is often called a **teedle board**.

My friends get on the seesaw with me.

Why can we not push ourselves off the ground?

My last stop is the slide. I climb up and slide down.

It is fun, but I want to go faster.

My friend says she can help me slide faster.

What can she do?

The first playground slides were made out of wood.

See what you have learned about pushes and pulls on the playground.

What do you push at the playground?
What do you pull at the playground?

KEY WORDS

Research has shown that as much as 65 percent of all written material published in English is made up of 300 words. These 300 words cannot be taught using pictures or learned by sounding them out. They must be recognized by sight. This book contains 72 common sight words to help young readers improve their reading fluency and comprehension. This book also teaches young readers several important content words, such as proper nouns. These words are paired with pictures to aid in learning and improve understanding.

Page	Sight Words First Appearance
4	and, away, day, every, from, move, them, things, to, us, we
5	about, are, can, I, learn, my, of, play, the, while
6	down, first, on, our, stop
7	back, begins, go, it, one
8	do, he, make, tell, that, want, what
9	at, for, have, people, sets, years
10	next
11	around, so
12	get, off
13	how, they
14	much, up
15	a, come, his, in, is, often, side
16	me, with
17	not, why
18	last
19	but
20	help, says, she
21	made, out, were

Page	Content Words First Appearance
5	friends, playground, pulls, pushes, science
6	swings
10	roundabout
14	seesaws
15	Massachusetts, teedle board
17	ground
18	slide
21	wood

Published by Smartbook Media Inc.
14 Penn Plaza, 9th Floor New York, NY 10122
Website: www.openlightbox.com

Library of Congress Control Number: 2020014163

ISBN 978-1-5105-5464-1 (hardcover)
ISBN 978-1-5105-5465-8 (multi-user eBook)

Printed in Guangzhou, China
1 2 3 4 5 6 7 8 9 0 24 23 22 21 20

052020
110819

Designer: Ana María Vidal Project Coordinator: Ryan Smith

The publisher acknowledges Getty Images, iStock, Dreamstime, and Shutterstock as the primary image suppliers for this title.